Agrivoltaics

Discover the Future of Sustainable Farming

I0490682

By Hugh Webb

Disclaimer:

The information contained in this book is for general information purposes only. And while we endeavor to keep the information up to date and correct, we make no representations or warranties of any kind, express or implied, about the completeness, accuracy, reliability, suitability or availability with respect to the book or the information, products, services, or related graphics contained in the book for any purpose. Any reliance you place on such information is therefore strictly at your own risk.

In no event will we be liable for any loss or damage including without limitation, indirect or consequential loss or damage, or any loss or damage whatsoever arising from loss of data or profits arising out of, or in connection with, the use of this book.

Table Of Contents

Call to action for further research and implementation of Agrivoltaics

Chapter 1: Definition of Agrivoltaics

Agrivoltaics is a cutting-edge technology that integrates agriculture and photovoltaic (PV) systems to produce both food and energy. This innovative approach merges two essential industries - agriculture and renewable energy - to create a sustainable and efficient solution for meeting the growing demand for both food and energy.

Agrivoltaics systems are typically installed over agricultural land, providing shading for crops and livestock while also generating clean electricity. The combination of agriculture and PV systems not only helps to maximize land usage, but it also creates a mutually beneficial relationship between the two. The shading provided by the PV panels creates a cooler environment for crops and livestock, reducing water usage and increasing yield. At the same time, the PV panels generate electricity for the farm or for the wider grid.

The key objective of Agrivoltaics is to create a harmonious and sustainable relationship between food production and energy generation. This innovative technology has the potential to address some of the world's most pressing issues such as food security, energy efficiency, and environmental protection.

In conclusion, Agrivoltaics is a promising solution that combines agriculture and photovoltaic technology to meet the growing demand for food and energy. By utilizing both traditional and modern approaches, Agrivoltaics provides a sustainable and efficient way of maximizing land usage and promoting a greener and more sustainable future.

Chapter 2: Importance of Agrivoltaics in Today's World

The world is facing a number of critical challenges, including food security, energy efficiency, and environmental protection. Agrivoltaics is a promising solution that addresses these challenges by combining agriculture and photovoltaic (PV) systems to produce both food and energy.

Food Security: With the global population projected to reach 9.7 billion by 2050, there is increasing pressure to increase food production to meet the growing demand. Agrivoltaics provides a solution by maximizing land usage and increasing food production through the shading effect provided by the PV panels. The technology also helps to conserve water and improve the overall health of crops and livestock.

Energy Efficiency: The world is dependent on non-renewable energy sources, which are finite and harmful to the environment. Agrivoltaics provides a sustainable alternative by generating clean, renewable energy through photovoltaic technology. The technology reduces reliance on non-renewable energy sources and helps to mitigate the impacts of climate change.

Environmental Protection: The use of non-renewable energy sources and traditional agriculture practices contribute to a range of environmental problems, including greenhouse gas emissions, land degradation, and water scarcity. Agrivoltaics provides a more sustainable solution that reduces the impact of agriculture on the environment while also generating clean, renewable energy.

In conclusion, Agrivoltaics is a crucial technology in today's world that addresses the challenges of food security, energy efficiency, and environmental protection. The integration of agriculture and photovoltaic technology provides a sustainable and efficient solution for meeting the growing demand for both food and energy.

Chapter 3: Purpose of the Book

The purpose of this book is to provide a comprehensive overview of Agrivoltaics and its applications in today's world. The book aims to educate and inform readers about this innovative technology and its benefits, and to encourage further research and implementation.

The book will explore the fundamentals of Agrivoltaics, including the history and evolution of the technology, the design of Agri-PV systems, and the benefits and limitations of Agrivoltaics. The book will also examine real-world Agrivoltaics projects and case studies, showcasing the applications of the technology in agriculture and livestock farming, greenhouses, and vertical farming.

In addition, the book will delve into the economic and environmental impacts of Agrivoltaics, highlighting the economic benefits and environmental benefits of this innovative technology. The book will also address the social and political factors that affect the growth and implementation of Agrivoltaics.

Finally, the book will look to the future of Agrivoltaics, examining the technological advancements, government policies and regulations, and the challenges and opportunities for growth in this exciting field.

The ultimate goal of this book is to promote understanding and awareness of Agrivoltaics and its potential to shape a more sustainable and efficient future. Whether you are a student, researcher, policy maker, or simply interested in the topic, this book will provide you with valuable information and insights into this important technology.

Chapter 4: Origin of Agrivoltaics

The concept of Agrivoltaics, or the integration of agriculture and photovoltaic (PV) systems, has its roots in the late 20th century. The idea of combining agriculture and energy generation was first proposed as a way to maximize land usage and address the growing demand for food and energy.

One of the earliest examples of Agrivoltaics was in the 1970s, when scientists in the United States began experimenting with using PV panels to provide shading for crops in greenhouses. The shading effect provided by the PV panels created a cooler environment for crops, reducing water usage and increasing yield. This was the first step in the development of Agrivoltaics as a technology.

In the following decades, the concept of Agrivoltaics evolved and expanded, and the technology was gradually adopted in other countries around the world. The integration of PV systems and agriculture became increasingly sophisticated, and the benefits of the technology became more widely recognized.

In recent years, the growth of the renewable energy industry and the increasing global demand for food and energy have fueled the development of Agrivoltaics. The technology is now seen as a promising solution to a number of critical challenges facing the world, including food security, energy efficiency, and environmental protection.

In conclusion, the origin of Agrivoltaics can be traced back to the 1970s, when scientists first began experimenting with using PV panels to provide shading for crops in greenhouses. Today, the technology has evolved and expanded, and is widely recognized as a promising solution to some of the world's most pressing challenges.

Chapter 5: Early Development of Agrivoltaics

The early development of Agrivoltaics can be traced back to the 1970s, when scientists first began experimenting with using photovoltaic (PV) systems to provide shading for crops in greenhouses. This early experimentation demonstrated the potential benefits of combining agriculture and energy generation, and paved the way for the development of Agrivoltaics as a technology.

One of the key benefits of using PV panels for shading in greenhouses was the reduction in water usage, as the shading effect provided by the PV panels created a cooler environment for crops. This reduced the need for irrigation and improved the overall health of crops.

In the following decades, the concept of Agrivoltaics evolved and expanded, and the technology was gradually adopted in other countries around the world. The first large-scale Agrivoltaics projects were developed in Europe and Asia, where the benefits of the technology were widely recognized and embraced.

Despite the growing recognition of the benefits of Agrivoltaics, the technology faced several challenges in its early development. The high cost of PV systems and the lack of government support and incentives were major barriers to the growth and adoption of Agrivoltaics. Additionally, the technical complexity of integrating agriculture and energy generation was a challenge for many early adopters.

Despite these challenges, the early development of Agrivoltaics demonstrated the potential of the technology to address some of the world's most pressing challenges, including food security, energy efficiency, and environmental protection. This early work laid the foundation for the continued growth and development of Agrivoltaics in the years to come.

Chapter 6: Modern Developments in Agrivoltaics

Since its early beginnings in the 1970s, the field of Agrivoltaics has experienced significant growth and development. The technology has evolved and expanded, and today it is seen as a promising solution to a number of critical challenges facing the world.

One of the key drivers of modern Agrivoltaics is the growing demand for renewable energy. As the world moves towards a more sustainable energy mix, Agrivoltaics has emerged as an attractive solution for harnessing solar energy and generating electricity on agricultural land. The integration of PV systems and agriculture allows for the efficient use of land and resources, and helps to address the growing demand for food and energy.

In recent years, the technology of Agrivoltaics has advanced significantly, with the development of new materials and components that make PV systems more efficient and cost-effective. The use of advanced control systems and monitoring technologies has also helped to optimize the performance of Agrivoltaics systems, improving their efficiency and effectiveness.

Governments around the world have also recognized the benefits of Agrivoltaics, and have introduced a range of policies and incentives to support its development and growth. These policies have helped to increase investment in the technology, and to accelerate its adoption and deployment.

In addition to its potential for energy generation, modern Agrivoltaics is also seen as a promising solution for improving food security and enhancing the sustainability of agriculture. The integration of agriculture and energy generation can help to reduce water usage and improve the overall health of crops, and can also provide new opportunities for farmers and rural communities.

In conclusion, modern developments in Agrivoltaics have transformed the technology from its early beginnings into a promising solution for addressing some of the world's most pressing challenges. The continued growth and development of Agrivoltaics is expected to drive further advances in the technology and to expand its impact on the world.

Chapter 7: Understanding Photovoltaic (PV) Technology

Photovoltaic (PV) technology is a critical component of Agrivoltaics and plays a vital role in the conversion of solar energy into electrical energy. Understanding the basic principles and components of PV technology is essential for fully appreciating the potential of Agrivoltaics and its applications.

At the heart of PV technology is the photovoltaic cell, which is the basic building block of PV systems. A photovoltaic cell is made up of two layers of semiconductor material, typically silicon, and when exposed to sunlight, generates a flow of electrons. This flow of electrons can be harnessed to generate electrical energy, which can be used to power homes, businesses, and other applications.

PV cells are combined into arrays, which are connected to form a complete PV system. The arrays are typically mounted on rooftops or other structures, and are designed to optimize their exposure to sunlight. The PV system also includes an inverter, which converts the direct current (DC) generated by the PV cells into alternating current (AC), which can be used to power homes and businesses.

One of the key advantages of PV technology is its scalability, which allows it to be used for a wide range of applications, from small-scale residential installations to large-scale commercial and utility-scale projects. Additionally, PV technology is modular and can be easily integrated into other systems, such as buildings and grid networks.

Another important advantage of PV technology is its environmental benefits. PV systems do not produce any emissions or pollutants, and generate electricity using a clean, renewable source of energy. This makes them an attractive solution for addressing the challenges of climate change and environmental degradation.

In conclusion, PV technology is a critical component of Agrivoltaics and plays a vital role in the conversion of solar energy into electrical energy. Understanding the basic principles and components of PV technology is essential for fully appreciating the potential of Agrivoltaics and its applications.

Chapter 8: Agri-PV System Design

Designing an Agri-PV system involves the integration of photovoltaic (PV) technology with agricultural practices and land use. The goal of Agri-PV system design is to create a system that is both effective in generating electrical energy and supportive of agricultural production.

The design process for an Agri-PV system typically begins with a site assessment, which involves evaluating the suitability of the land and the potential for integrating PV technology with agriculture. Factors such as soil quality, topography, and available sunlight are all considered in the site assessment.

Once the site has been assessed, the design team will begin to develop a conceptual design for the Agri-PV system. This will involve determining the optimal orientation and tilt for the PV arrays, as well as the size and capacity of the system. The design will also take into account the needs of the crops and the requirements of the local environment.

The design process also involves the selection of components and materials for the Agri-PV system. This includes the choice of photovoltaic cells, inverters, mounting systems, and other components that are critical to the performance of the system.

In addition to the technical considerations, the design process also takes into account the social and economic factors associated with the deployment of an Agri-PV system. This includes assessing the potential impact of the system on the local community, as well as the potential benefits in terms of job creation and economic development.

Once the conceptual design has been developed, the design team will conduct a feasibility study to evaluate the economic viability of the project. This will involve assessing the costs associated with the construction and operation of the system, as well as the potential revenue streams from the sale of electricity.

In conclusion, the design of an Agri-PV system involves the integration of photovoltaic technology with agricultural practices and land use. The design process is an iterative process that takes into account a range of technical, social, and economic considerations, and is critical to the success of an Agri-PV project.

Chapter 9: Benefits and Limitations of Agrivoltaics

Agrivoltaics is a relatively new and evolving field that combines photovoltaic (PV) technology with agriculture, offering a range of benefits and limitations. In this chapter, we will explore some of the key benefits and limitations of Agrivoltaics.

Benefits of Agrivoltaics:

Increased Energy Generation: The integration of PV technology with agriculture can result in increased energy generation, as the PV arrays can be placed over farmland, thereby maximizing the use of available land and sunlight.

Improved Crop Performance: Agrivoltaics can provide shade and protection to crops, improving their performance and reducing water loss.

Dual Land Use: Agrivoltaics allows for the simultaneous use of land for both energy production and agriculture, making the most efficient use of available resources.

Environmental Benefits: Agrivoltaics generates clean and renewable energy, reducing greenhouse gas emissions and contributing to the fight against climate change.

Economic Benefits: Agrivoltaics can provide economic benefits by creating new revenue streams from the sale of electricity, as well as through the creation of jobs in the installation and maintenance of PV systems.

Limitations of Agrivoltaics:

Cost: The initial costs associated with the installation of an Agri-PV system can be high, making it a less attractive option for some farmers and land owners.

Technical Challenges: The integration of PV technology with agriculture can be challenging, requiring a deep understanding of both fields to ensure optimal performance.

Maintenance: Agri-PV systems require regular maintenance to ensure that they are functioning properly and producing energy efficiently.

Availability of Land: The availability of suitable land for Agri-PV systems may be limited in some areas, reducing the potential for widespread adoption of the technology.

In conclusion, Agrivoltaics offers a range of benefits, including increased energy generation, improved crop performance, dual land use, environmental benefits, and economic benefits. However, it also faces a number of limitations, including cost, technical challenges, maintenance requirements, and limited availability of suitable land. Despite these limitations, the potential benefits of Agrivoltaics make it an exciting and promising field with significant potential for growth and development.

Chapter 10: Agriculture and Livestock Farming in Agrivoltaics

Agriculture and livestock farming play a significant role in the implementation of Agrivoltaics. The integration of photovoltaic (PV) technology with agriculture provides a range of benefits for both the energy and agriculture sectors, making it an important area of focus for both industries.

Agriculture and Agrivoltaics:

Improved Crop Performance: Agrivoltaics can provide shade and protection to crops, improving their performance and reducing water loss. The PV arrays can also protect crops from wind and hail damage, reducing the risk of crop losses.

Increased Land Use Efficiency: By integrating PV technology with agriculture, the available land can be used more efficiently, maximizing the use of resources and increasing productivity.

Environmental Benefits: Agrivoltaics generates clean and renewable energy, reducing greenhouse gas emissions and contributing to the fight against climate change. Additionally, the integration of PV technology with agriculture can reduce the need for energy-intensive farming practices, further reducing emissions and promoting sustainable agriculture.

Livestock Farming and Agrivoltaics:

Improved Livestock Health: Agrivoltaics can provide shade and protection for livestock, improving their health and reducing the risk of heat-related illnesses.

Increased Land Use Efficiency: Livestock farming and Agrivoltaics can also be integrated, allowing for the efficient use of available land and resources.

In conclusion, the integration of photovoltaic technology with agriculture and livestock farming provides a range of benefits, including improved crop and livestock health, increased land use efficiency, and environmental benefits. This makes Agrivoltaics an important area of focus for both the energy and agriculture industries, offering significant potential for growth and development.

Chapter 11: Greenhouses and Vertical Farming in Agrivoltaics

Greenhouses and vertical farms are two key areas where the integration of photovoltaic (PV) technology with agriculture is being utilized to provide significant benefits. Both greenhouses and vertical farms provide unique opportunities for Agrivoltaics to improve the efficiency of agriculture and support sustainable growth.

Greenhouses and Agrivoltaics:

Improved Climate Control: Greenhouses can benefit from the integration of PV technology as the panels can provide shade and control the temperature within the greenhouse, improving the growing conditions for crops.

Increased Energy Efficiency: By generating clean and renewable energy within the greenhouse, energy costs can be reduced, increasing efficiency and reducing greenhouse gas emissions.

Increased Productivity: The integration of PV technology with greenhouses can provide optimal growing conditions for crops, resulting in increased productivity and improved yields.

Vertical Farming and Agrivoltaics:

Space-Saving: Vertical farms can be designed to incorporate PV technology, allowing for efficient use of space and resources.

Optimal Growing Conditions: Vertical farms can benefit from the integration of PV technology, providing optimal growing conditions and reducing the risk of crop losses.

Increased Energy Efficiency: Vertical farms can generate clean and renewable energy, reducing energy costs and greenhouse gas emissions.

In conclusion, the integration of photovoltaic technology with greenhouses and vertical farms provides significant benefits, including improved climate control, increased energy efficiency, and increased productivity. This makes Agrivoltaics an important area of focus for the agriculture industry, offering significant potential for growth and development.

Chapter 12: Real-World Agrivoltaics Projects and Case Studies

Agrivoltaics is rapidly gaining traction as a viable solution for integrating photovoltaic (PV) technology with agriculture. Real-world projects and case studies provide valuable insights into the benefits and challenges of implementing Agrivoltaics.

Agri-PV Farm in France: This Agri-PV farm in France is a leading example of how Agrivoltaics can be integrated with agriculture, providing significant benefits for both the energy and agriculture sectors. The project features a combination of crops, livestock, and PV technology, demonstrating the versatility of Agrivoltaics.

Vertical Farm in Japan: A vertical farm in Japan is utilizing Agrivoltaics to provide optimal growing conditions for crops, reducing the risk of crop losses and improving yields. The integration of PV technology with the vertical farm has also reduced energy costs and greenhouse gas emissions, promoting sustainability.

Greenhouse in the Netherlands: A greenhouse in the Netherlands is utilizing Agrivoltaics to provide optimal growing conditions for crops, improving yields and reducing energy costs. The integration of PV technology has also provided climate control within the greenhouse, improving the growing conditions for crops.

Livestock Farm in Germany: A livestock farm in Germany is utilizing Agrivoltaics to provide shade and protection for livestock, improving their health and reducing the risk of heat-related illnesses. The integration of PV technology has also reduced energy costs and greenhouse gas emissions, promoting sustainability.

In conclusion, real-world Agrivoltaics projects and case studies provide valuable insights into the benefits and challenges of implementing Agrivoltaics. These projects demonstrate the versatility and potential of this innovative solution, offering significant potential for growth and development in the agriculture industry.

Chapter 13: Economic and Environmental Impacts of Agrivoltaics

The integration of photovoltaic (PV) technology with agriculture, known as Agrivoltaics, has significant economic and environmental impacts. Understanding these impacts is essential for the growth and development of the Agrivoltaics industry.

Economic Impacts:

Reduced Energy Costs: The integration of PV technology with agriculture can reduce energy costs, providing significant economic benefits.

Improved Yields: Agrivoltaics can provide optimal growing conditions for crops, improving yields and increasing productivity.

Increased Revenue: The integration of PV technology with agriculture can generate additional revenue through the sale of clean and renewable energy.

Environmental Impacts:

Reduced Greenhouse Gas Emissions: Agrivoltaics can reduce greenhouse gas emissions, promoting sustainability and mitigating the impacts of climate change.

Improved Soil Health: The integration of PV technology with agriculture can improve soil health, supporting the long-term viability of agriculture.

Increased Biodiversity: Agrivoltaics can support increased biodiversity, promoting healthy ecosystems and preserving wildlife habitats.

In conclusion, the economic and environmental impacts of Agrivoltaics are significant and provide compelling reasons for continued growth and development of this innovative solution. The integration of PV technology with agriculture offers significant potential for economic and environmental benefits, making it an important area of focus for the agriculture industry.

Chapter 14: Economic Benefits of Agrivoltaics

The integration of photovoltaic (PV) technology with agriculture, known as Agrivoltaics, offers significant economic benefits. Understanding these benefits is essential for the growth and development of the Agrivoltaics industry.

Reduced Energy Costs: The integration of PV technology with agriculture can reduce energy costs, providing significant economic benefits. The use of clean and renewable energy can reduce reliance on traditional energy sources, lowering energy bills and increasing profitability.

Improved Yields: Agrivoltaics can provide optimal growing conditions for crops, improving yields and increasing productivity. By providing shade and climate control, Agrivoltaics can reduce the risk of crop losses and improve the quality of crops, increasing revenue.

Increased Revenue: The integration of PV technology with agriculture can generate additional revenue through the sale of clean and renewable energy. The generation of renewable energy can provide a new source of income, supporting the long-term viability of the agriculture industry.

Access to Government Incentives: Agrivoltaics can provide access to government incentives and subsidies, reducing the cost of implementation and increasing profitability.

In conclusion, the economic benefits of Agrivoltaics are significant and provide compelling reasons for continued growth and development of this innovative solution. The integration of PV technology with agriculture offers significant potential for economic benefits, making it an important area of focus for the agriculture industry.

Chapter 15: Environmental Benefits of Agrivoltaics

The integration of photovoltaic (PV) technology with agriculture, known as Agrivoltaics, offers significant environmental benefits. Understanding these benefits is essential for the growth and development of the Agrivoltaics industry.

Reduced Greenhouse Gas Emissions: Agrivoltaics can reduce greenhouse gas emissions, promoting sustainability and mitigating the impacts of climate change. The use of clean and renewable energy can reduce reliance on traditional energy sources, reducing carbon emissions and improving air quality.

Improved Soil Health: The integration of PV technology with agriculture can improve soil health, supporting the long-term viability of agriculture. By providing shade and reducing the need for water irrigation, Agrivoltaics can improve soil moisture levels and fertility, promoting healthy plant growth.

Increased Biodiversity: Agrivoltaics can support increased biodiversity, promoting healthy ecosystems and preserving wildlife habitats. The integration of PV technology with agriculture can provide habitat for wildlife, reducing the need for additional land development and preserving natural ecosystems.

Enhanced Land Use Efficiency: Agrivoltaics can enhance land use efficiency, reducing the amount of land needed for agriculture and energy production. By integrating PV technology with agriculture, the same land can be used for multiple purposes, increasing land use efficiency and reducing the need for additional land development.

In conclusion, the environmental benefits of Agrivoltaics are significant and provide compelling reasons for continued growth and development of this innovative solution. The integration of PV technology with agriculture offers significant potential for environmental benefits, making it an important area of focus for the agriculture industry.

Chapter 16: Social and Political Factors affecting Agrivoltaics

The growth and development of Agrivoltaics, the integration of photovoltaic (PV) technology with agriculture, is influenced by a range of social and political factors. Understanding these factors is essential for the continued growth and development of the Agrivoltaics industry.

Public Awareness: Public awareness and understanding of Agrivoltaics can impact its growth and development. Increasing public awareness of the benefits of Agrivoltaics, including economic, environmental, and social benefits, can support the growth of the industry.

Government Support: Government support and funding can play a significant role in the growth and development of Agrivoltaics. Government policies and incentives, such as subsidies and tax credits, can reduce the cost of implementation and promote the growth of the industry.

Technological Advancements: Technological advancements can impact the growth and development of Agrivoltaics. The availability of new technologies, including advanced PV systems and efficient energy storage systems, can support the growth of the industry and improve the viability of Agrivoltaics solutions.

Social and Political Stability: Social and political stability can impact the growth and development of Agrivoltaics. Political stability and economic stability can support the growth of the industry, while political unrest and economic instability can negatively impact the growth of the industry.

In conclusion, social and political factors can significantly impact the growth and development of Agrivoltaics. Understanding these factors is essential for the continued growth and development of this innovative solution, and for promoting its growth and viability in the long-term.

Chapter 17: Future of Agrivoltaics

Agrivoltaics, the integration of photovoltaic (PV) technology with agriculture, has the potential to revolutionize the way we produce and consume energy. With continued advancements in technology and growing public awareness of the benefits of Agrivoltaics, the future of this innovative solution is bright.

Technological Advancements: Technological advancements in the field of Agrivoltaics will continue to drive its growth and development. The development of new, more efficient PV systems and energy storage systems will further improve the viability of Agrivoltaics solutions and increase their adoption.

Growing Demand for Renewable Energy: The growing demand for renewable energy, driven by concerns about climate change and energy security, will support the growth of Agrivoltaics. As the world looks to transition away from fossil fuels, Agrivoltaics will play an increasingly important role in meeting our energy needs.

Increased Adoption: The adoption of Agrivoltaics will continue to grow as more people become aware of its benefits and as technological advancements make it more viable. The development of new business models, such as community-based Agrivoltaics, will further drive its growth and increase its adoption.

Government Support: Government support and funding will play a critical role in the future of Agrivoltaics. Government policies and incentives, such as subsidies and tax credits, will continue to support the growth of the industry and increase the adoption of Agrivoltaics solutions.

In conclusion, the future of Agrivoltaics is bright and holds immense potential for the production and consumption of renewable energy. The continued growth and development of this innovative solution will play a critical role in shaping our energy future and addressing the challenges of climate change and energy security.

Chapter 18: Technological Advancements in Agrivoltaics

The integration of photovoltaic (PV) technology with agriculture, known as Agrivoltaics, has seen rapid advancements in recent years. The development of new, more efficient PV systems and energy storage systems, has improved the viability of Agrivoltaics solutions and increased their adoption. In this chapter, we will explore some of the key technological advancements in Agrivoltaics.

Improved PV Systems: One of the key technological advancements in Agrivoltaics is the development of improved PV systems. New PV systems are more efficient, durable and cost-effective than their predecessors, making them more viable for use in Agrivoltaics.

Energy Storage Systems: Another important advancement in Agrivoltaics is the development of energy storage systems. Energy storage systems allow for the storage of excess energy generated by PV systems, making Agrivoltaics solutions more reliable and sustainable.

Smart Agriculture: The integration of Agrivoltaics with smart agriculture technologies is another area of growth. Smart agriculture technologies, such as precision agriculture and vertical farming, can be integrated with Agrivoltaics to improve the efficiency and productivity of agricultural operations.

Remote Monitoring and Management: The development of remote monitoring and management technologies has improved the viability of Agrivoltaics solutions. These technologies allow for the monitoring and management of Agrivoltaics systems from a remote location, reducing the need for on-site maintenance and increasing the reliability of the systems.

Community-Based Agrivoltaics: The development of community-based Agrivoltaics is another area of growth in the field. Community-based Agrivoltaics allows for the sharing of resources and expertise, making it more accessible and cost-effective for small-scale farmers and rural communities.

In conclusion, the technological advancements in Agrivoltaics are driving its growth and increasing its viability as a renewable energy solution. As technology continues to evolve, Agrivoltaics will play an increasingly important role in shaping our energy future.

Chapter 19: Challenges and Opportunities for Agrivoltaics Growth

The growth of Agrivoltaics, the integration of photovoltaic (PV) technology with agriculture, presents both challenges and opportunities. In this chapter, we will examine the challenges and opportunities facing the growth of Agrivoltaics.

Technical Challenges: One of the major challenges facing the growth of Agrivoltaics is the technical complexity of integrating PV systems with agriculture. This includes issues with shading, energy storage, and system maintenance.

Economic Challenges: The high cost of PV systems and energy storage systems is another challenge facing the growth of Agrivoltaics. This can make it difficult for small-scale farmers and rural communities to adopt Agrivoltaics solutions.

Regulatory Challenges: The regulatory environment is another challenge facing the growth of Agrivoltaics. This includes issues with permitting, zoning, and interconnection with the grid.

Market Challenges: The lack of awareness and understanding of Agrivoltaics among potential users is another challenge facing its growth. This includes a lack of information on the benefits and limitations of Agrivoltaics and limited access to financing.

Opportunities for Growth: Despite these challenges, there are also many opportunities for growth in the field of Agrivoltaics. This includes the increasing demand for renewable energy, the development of new, more efficient PV systems, and the growth of smart agriculture technologies.

Government Support: Government support for Agrivoltaics, through funding, research and development programs, and incentives, can also help to drive its growth.

In conclusion, the growth of Agrivoltaics presents both challenges and opportunities. Addressing these challenges and capitalizing on the opportunities will be essential for the continued growth of Agrivoltaics and its role as a renewable energy solution.

Chapter 20: Summary of Key Points

In this book, we have explored the topic of Agrivoltaics, the integration of photovoltaic (PV) technology with agriculture. We have covered the definition, importance, origin, development, and future of Agrivoltaics. We have also explored the benefits and limitations of this innovative technology, including its economic and environmental impacts.

Here are some of the key points covered in the book:

Definition of Agrivoltaics: Agrivoltaics refers to the integration of photovoltaic (PV) technology with agriculture to provide renewable energy and increase food production.

Importance of Agrivoltaics: Agrivoltaics has the potential to provide a sustainable solution for both energy and food production, meeting the growing demand for both in today's world.

Origin of Agrivoltaics: The concept of Agrivoltaics has its roots in the energy crisis of the 1970s, when researchers began exploring ways to integrate renewable energy sources with agriculture.

Modern Developments in Agrivoltaics: Today, Agrivoltaics has advanced significantly, with the development of new, more efficient PV systems and the integration of smart agriculture technologies.

Benefits and Limitations: Agrivoltaics provides a range of benefits, including increased energy production, improved crop yields, and reduced energy costs. However, there are also limitations, including high costs, shading issues, and technical complexity.

Economic and Environmental Impacts: Agrivoltaics has the potential to provide significant economic and environmental benefits, including job creation, reduced carbon emissions, and increased food security.

Future of Agrivoltaics: The future of Agrivoltaics is bright, with increasing demand for renewable energy and the development of new, more efficient technologies.

In conclusion, this book provides a comprehensive overview of Agrivoltaics and its potential to provide a sustainable solution for both energy and food production. By understanding the key points covered in this book, we can help to drive the growth of this innovative technology and move towards a more sustainable future.

Chapter 21: Final Thoughts and Recommendations

Throughout this book, we have explored the topic of Agrivoltaics and the integration of photovoltaic (PV) technology with agriculture. We have covered the definition, importance, origin, development, benefits, and limitations of Agrivoltaics, as well as its future and technological advancements.

In this final chapter, we will summarize the key findings and provide some recommendations for the future of Agrivoltaics.

Key Findings:
Agrivoltaics has the potential to provide a sustainable solution for both energy and food production, meeting the growing demand for both in today's world.
The benefits of Agrivoltaics include increased energy production, improved crop yields, reduced energy costs, and reduced carbon emissions.
There are also limitations to Agrivoltaics, including high costs, shading issues, and technical complexity.
The future of Agrivoltaics is bright, with increasing demand for renewable energy and the development of new, more efficient technologies.
Recommendations:
Governments should provide incentives and support for the development and implementation of Agrivoltaics projects.
The private sector should invest in R&D to develop more efficient and cost-effective Agrivoltaic systems.
Agricultural organizations should collaborate with energy companies to explore the potential of Agrivoltaics and its benefits for both industries.
The public should be educated on the benefits of Agrivoltaics and how it can help to move towards a more sustainable future.

In conclusion, Agrivoltaics has the potential to provide a sustainable solution for both energy and food production. By implementing the recommendations outlined in this chapter, we can help to drive the growth of this innovative technology and move towards a more sustainable future. We hope that this book has provided valuable insights into the world of Agrivoltaics and its potential to shape our future.

Chapter 22: Call to Action for Further Research and Implementation of Agrivoltaics

In this final chapter, we present a call to action for further research and implementation of Agrivoltaics. The integration of photovoltaic (PV) technology with agriculture has the potential to provide a sustainable solution for both energy and food production, meeting the growing demand for both in today's world.

Further Research:
There is a need for further research to fully understand the potential benefits and limitations of Agrivoltaics.
R&D is required to improve the efficiency and cost-effectiveness of Agrivoltaic systems, making them more accessible to a wider range of users.
Studies should be conducted to better understand the social and political factors that affect the implementation of Agrivoltaics, and how to address them.
Implementation:
Governments should provide incentives and support for the development and implementation of Agrivoltaics projects.
The private sector should invest in Agrivoltaics and work with governments and agricultural organizations to promote its growth.
Agricultural organizations should work to integrate Agrivoltaics into their operations, taking advantage of the benefits it offers.
The public should be made aware of the benefits of Agrivoltaics and encouraged to support its implementation.

In conclusion, there is a clear need for further research and implementation of Agrivoltaics. By taking action on the recommendations outlined in this chapter, we can help to drive the growth of this innovative technology and move towards a more sustainable future. We hope that this book has inspired you to take an active role in supporting the development of Agrivoltaics and making it a reality.

Epilogue

As the world continues to grapple with the effects of change, sustainable agriculture has never been more important. Agrivoltaics provides a powerful solution to this global challenge, offering a way to reduce carbon emissions, conserve water resources, and increase crop yields.

But agrivoltaics is not just a theoretical concept. It is a growing movement of farmers, researchers, and innovators who are working to create a more sustainable future for agriculture. By combining solar energy production with agricultural practices, agrivoltaics offers a new paradigm for farming that is both profitable and environmentally friendly.

As we look to the future, the potential of agrivoltaics is truly exciting. From rooftop gardens to large-scale solar farms, the possibilities for combining solar energy and agriculture are endless. By embracing agrivoltaics, we can create a world where food production is sustainable, energy is renewable, and the planet is thriving.

So let us continue to explore and innovate in the field of agrivoltaics. Let us work together to create a brighter and more sustainable future for ourselves and for generations to come.